Splitting Colonies for the Small-Scale Beekeeper
David MacFawn

ISBN: 978-1-914934-44-5

Published by Northern Bee Books 2022
Northern Bee Books, Scout Bottom Farm, Mytholmroyd,
Hebden Bridge HX7 5JS (UK).
www.northernbeebooks.co.uk
+44 (0) 1422 882751.

Book design by www.SiPat.co.uk

Splitting Colonies
for the Small-Scale Beekeeper

David MacFawn

Contents

Figures

Introduction

There are multiple ways to increase colonies and control swarming. Splitting colonies is one method to help the beekeeper. Walk-away splits (where you make sure each split half contains less than three-day-old larvae, honey, and pollen and let the queenless split raise another queen) or installing queen cells or a mated queen are additional ways to split a colony for a colony number increase. Demaree (see later in article for a more detailed explanation) is another technique to simulate a split but keep all the worker bees in the same hive. If you are looking for the queen, she is usually found on a frame with freshly laid eggs.

There is a saying in beekeeping: A poor genetic queen raised in an excellent environment is superior to a better genetic queen raised in a poor environment. We want superior genetic queens raised in an excellent environment. The definition of "superior genetics" may change over time. The queen's glands and body are better developed in an excellent environment.

Figure 1. The queen (red dot on thorax) yearly international queen marking colors placed by the beekeeper on the queen's thorax representing the year last digit age of the queen (1 & 6 white, 2 & 7 yellow, 3 & 8 red, 4 & 9 green, 5 & 0 blue).

Good queens have 150 to 180 ovarioles per two ovaries[1] and are large.

If the brood pattern is small or irregular it may indicate a poor queen. A good queen tends to lay eggs in the center of the centermost combs and radiate outward as she expands her egg-laying. The oldest brood will be toward the center and emerge first. An irregular brood pattern may also indicate hygienic bees—bees that remove larvae and pupa that have varroa or are diseased. A "normal" brood nest should be looked for in the spring with brood in the frame's center, typically a band of pollen, then honey in the corners. This brood nest configuration may expand over a brood chamber and a super.[2]

A generic timeline for splitting and spring buildup

Weeks to spring nectar flow	Event
	winter solstice/begin spring buildup
nine to ten weeks	start feeding if your goal is to split colonies
eight to nine weeks	henbit, maples, dandelions, willow bloom/early bloom in your area
four to five weeks	check for purple-eye drones and swarm cells
four to five weeks	check for swarming & do early splits or other swarm control measures
four to five weeks	earliest start for queen rearing
two to three weeks	start queen rearing when drones are walking on comb: conservative approach
0	spring flow starts

---

1   Hive & Honey Bee 2015, ISBN 978-0-915698-16-5 edited by Joe Graham, Dadant, page 161
2   Hive & Honey Bee 2015, ISBN 978-0-915698-16-5 edited by Joe Graham, Dadant, page 495

In the Columbia, South Carolina, area, colonies can be split at the end of February/early March at the earliest. Adding empty supers with frames will not relieve the congestion in the brood nest that usually causes swarming. Frames with brood and bees need to be removed in the congested brood nest. If colonies are fed sugar syrup mid to the end of January, most colonies will be ready to split from the end of February to the first of March at the earliest (approximately two brood cycles). Usually, only healthy colonies that are well provisioned with honey and pollen build up sufficiently to split. Maples bloom at the end of January to the first of February in the Columbia, South Carolina, area and are considered a major pollen source and a minor nectar source. The nectar flow usually starts around the first of April and continues through the first part of June in this region.

Colonies can be split consistently in South Carolina as early as the end of February if fed starting the second half to the end of January. The colony should be split when it starts warming consistently into the upper 30s°F. to lower 40s°F. (2°C. to 8°C.) at night. The split needs enough worker bees to cover the brood. When splitting, each split half should have eggs/less than three-day-old larvae, honey, and pollen. Each split-half should be fed.

The rule of thumb is a colony can be split when there are purple-eyed drone pupa. However, this needs to be qualified. The drone and queen time durations are:

- Fifteen-day-old purple-eye drone pupa with nine days left of the 24-day development time + 14 to 16 days maximum to sexually mature (actually seven to 14 days) equals approximately 23 to 25 days (or about three to three and a half weeks) to sexually mature queens.

- Sixteen-day development from egg plus four to seven days to sexually mature equals 20 to 23 days or about three weeks.[3]

Splitting the colony should be held off for three to four days after purple-eyed drone pupa are observed. The queen egg takes three days to mature into feedable larvae. This means, that after purple-eyed drones are observed, the beekeeper should wait three to four days to ensure the queen emerges when there are sexually mature drones. Also, there should be a multitude of purple-eyed drones before splitting, not just a few. You want to be on the mature-side average of the purple-eyed drone larvae. A more conservative view is to have adult drones walking around on combs.

---

3 ABC and XYZ of Bee Culture, 42nd edition, 2020, ISBN-13 978-0-9846915-3-1 page 460

Figure 2. Hive on February 16, 2022, in Bishopville, SC, configuration before splitting. The top deep contains a two-gallon pail feeder. (Photo courtesy: David MacFawn)

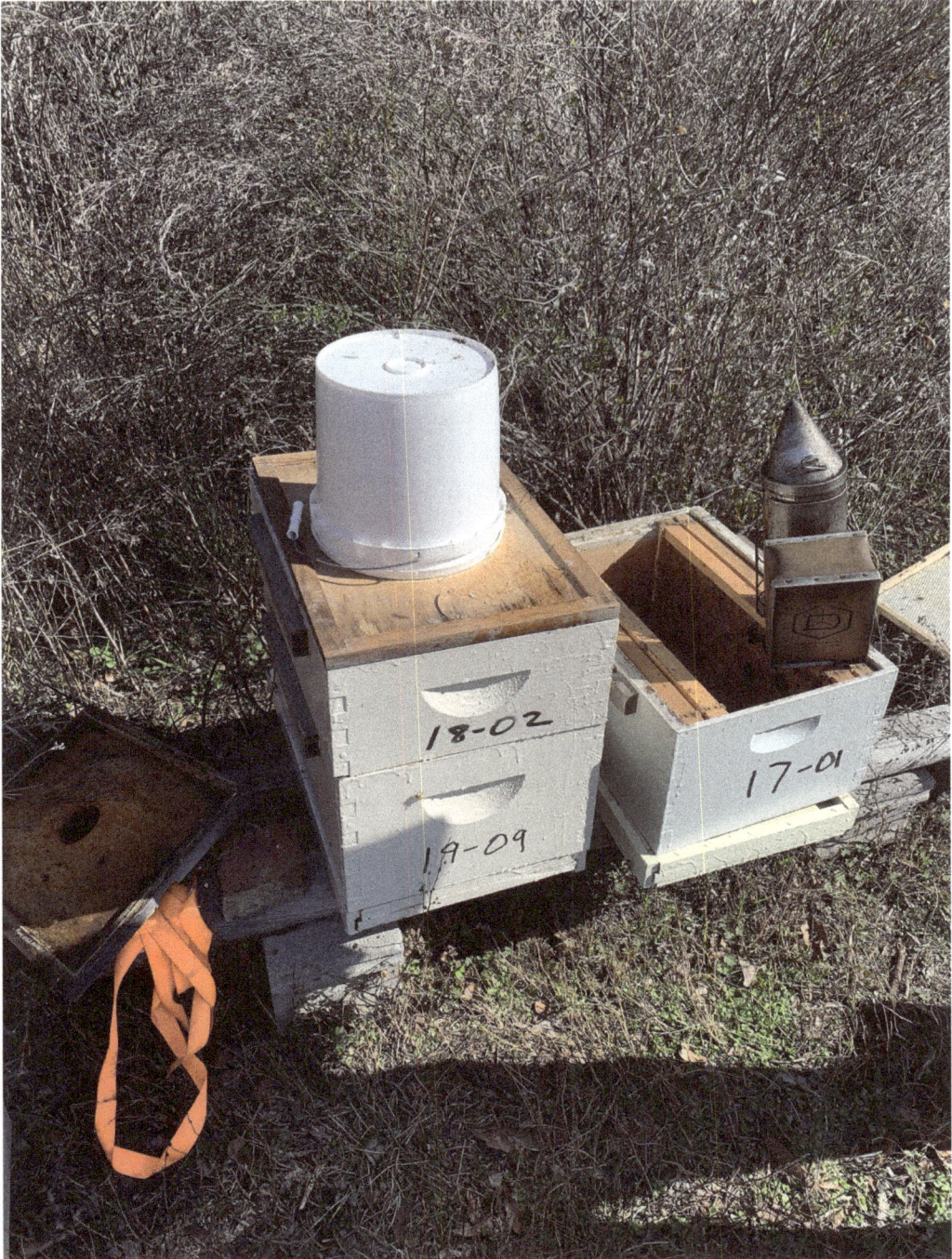

Figure 3. Note two-gallon pail feeder over porter bee escape hole in the inner cover. In South Carolina we can feed syrup year-round with success due to moderate weather. (Photo courtesy: David MacFawn)

Splitting and purple-eyed drone pupa occurs earliest (in January and February) in the deep south/Florida and progresses farther north as spring approaches in the country. In South Carolina, it occurs from the end of February into March. The beekeeper should consult their local bee association members to determine when splitting and thus purple-eyed drone pupa normally occurs.

If a walk-away split is made, where the bees raise their queen, it will take approximately 15 to 17 days for the queen to emerge, ten to 14 days to mate and start laying, 19 to 21 days for workers to emerge in a new split, and an additional 18 to 21 days to become field bees. In total that is approximately nine to ten weeks to produce field bees from a walk-away split. This maturation time will impact honey production. The old saying you can make either bees or honey is true. It should be noted in the southeast United States there is typically a dearth from mid-June until August with a weak autumn nectar flow. When you evaluate your colonies in the August/September time frame for overwintering, a colony split the previous spring may not be fully developed and may need feeding to get through winter. This is also why only a small percentage of swarms survive the first year. In South Carolina, if the beekeeper can find a summer nectar flow such as cotton, the splits can be developed before autumn. Colonies may need to be moved to this other nectar flow for their development.

When making a split, the split should be closely monitored for about three weeks to see if you need to add another frame of brood. Bees typically only live an average of five to six weeks during warm weather and it may take nine to ten weeks to obtain field bees. Splits should be fed sugar syrup as a safeguard with the understanding that during the nectar flow stored syrup may intermingle with collected nectar. This intermingling is considered adulterated honey. During warm weather in the southeast, there is usually plenty of pollen. Bees typically will stop using sugar syrup when fresh nectar is available. However, due to the reduced size of a split, and normal population losses, it is wise to offer sugar syrup until the split becomes a functional colony. If another frame of bees and brood is added from a strong non-split colony, and if the nectar flow is one-quarter to one-half complete, the brood that is removed from the strong non-split colony will affect honey production of the donor colony minimally. This is because the worker bee development time needed for any brood taken from the non-split donor colony has yet to complete its development into mature field bees. The worker bees will complete their development after the nectar flow is over. In South Carolina, the first half of June is an excellent time to split (right after the nectar flow). Colonies may also be split in August, but a mated queen and drawn comb are preferred for overwintering success. If Small Hive Beetles (SHB) are an issue in your area when splitting make sure you have enough bees to patrol the hive space. Place the split in a five-frame NUC, eight-frame hive depending on its size to make sure you have enough bees to patrol the hive space.

If a walk-away split is done at the end of February, it takes approximately three weeks to raise a queen, a week or so to mate and start laying, and another three weeks for the first workers to emerge. This puts the first workers emerging mid-April from the queenless split half, with the nectar flow starting around the first of April. Approximately another three weeks are required for the house worker bees to mature into field bees. Then the nectar-gathering field bees for the walk-away split will be ready early to mid-May. The nectar flow is over usually around the first of June, so the walk-away split has missed most of the spring nectar flow. This means a walk-away split will need to be fed during the summer dearth that starts mid-June and runs through until around the first of August. If the colony is taken to cotton fields for nectar that blooms mid-July through September or sourwood bloom in the higher mountain elevations feeding is unnecessary. It should be noted that sometimes an inferior queen may result from a walk-away split. A walk-away split queen can be evaluated and replaced if this occurs. Dr. David Tarpy's lab at North Carolina State University indicated from their queen research if the capped queen cells are culled at exactly five days after splitting, results in a reasonable queen. Often the bees will choose an older larva to produce a walk away queen. A walk-away split may be required in South Carolina at the end of February due to a lack of mated queen availability. The split-half with the original queen should continue to build up properly and often you will get a reasonable honey crop. Walk-away split results in the new queen split obtaining the genetics from the local area.

A generic walk-away split schedule is:

Event	Week
Split	0
Approximately 3 weeks to raise a queen	3 weeks
A week or so to mate and start laying	4 weeks
3 weeks for the first workers to emerge	7 weeks
Approximately 3 weeks are required for the house worker bees to mature into field bees	10 weeks

Figure 4. Outer cover on top of pail feeder chamber made with a sheet of OSB, insulation, and another sheet OSB for thermal mass. (Photo courtesy: David MacFawn)

When splitting a colony, I usually do not find the queen. I make sure both split halves have plenty of honey, pollen, and less than three-day-old larvae. This results in the split that does not have the old queen to start raising a queen and the worker brood from the original queen will emerge in approximately one to two weeks. The split-half without worker brood in approximately two weeks will be the queenless split. It takes a lot of time to find queens. I minimize my labor costs by not finding the queen. The colony should be monitored for laying worker activity. Usually, you have three to six weeks before a laying worker occurs.

One to two days after splitting, queen cell(s) can be installed in both splits. Queen cells are much cheaper than mated queens. This results in an approximate one-week head start on a walk-away split. A mated queen can also be installed in the split-half that does not have brood after approximately one and a half weeks.

Whether you do a walk-away split or use queen cells or mated queen or a Demaree depends on what your strategy is. A walk-away split and queen cells can be used to get half your genetics from the local area if you are interested in local area genetics. A mated queen, if not raised from local colonies, can be used to replace 100 percent of your colony genes from the queen mated area of the mated queen. A Demaree is when all the brood in the bottom brood chamber is moved above a queen excluder and a super with frames. The queen is left on a frame of drawn comb in the bottom brood chamber. In seven to ten days, the brood above the queen excluder is checked for queen cells and the queen cells are destroyed to inhibit swarming. The colony believes it has swarmed and you have retained all your field bees for honey production. After several weeks the Demaree colony can be recombined.

Figure 5. Note the number of bees on the inner cover. This colony was ready to split on February 16, 2022, in Bishopville, SC (purple-eyed drone pupae). (Photo courtesy: David MacFawn)

Figure 6. Strapping the split and getting it ready to move. (Photo courtesy: David MacFawn)

Figure 7. A mated queen in a Benton queen cage. (Photo courtesy: David MacFawn)

When selecting a frame of brood, you want capped brood with some eggs and larvae. Capped brood has completed its feeding stage which means the split does not need as many nurse bees. The eggs and larvae are an insurance policy in case the bees need to raise a new queen. Also, you want all different stages and time frames of brood so you have a continuous supply of bees. A minimum three-frame split is recommended with a frame of capped brood, eggs and larvae, honey, and pollen. A five-frame split is preferred with all stages of brood since it will build up quicker than a three-frame split.

Raising local queens is preferred over using queens imported from another region. Local queens have developed survival characteristics from the local area and are thought to have a survival edge over queens imported from another region. Beekeeping is best when it's local and local queens are very important. Raising a few queens can be done with the Miller Method (foundation cut to a sharp point), Alley Method, Doolittle Method (grafting), or the Hopkins Method (turning the frame of eggs/larvae on its side in a queenless colony).[4,5,6] The virgin queens should be placed in NUCs (nucleus colonies) and left until they are laying fertilized worker brood (for at least 28 days).

4   Graham, Joe M., The Hive and the Honey Bee, Dadant 2015 ISBN 978-0-91-915698-16-5.
5   Gilles Fert, Breeding Queens, published by O.P.I.D.A., ISBN 2 905851 11-2.
6   Roger A. Morse, Rearing Queen Honey Bees, 2nd Edition, Wicwas Press, ISBN 978-1-878075-05-5.

Several characteristics are often looked for in propagating queen stock. Disease resistance, overwintering ability, swarm tendencies, honey production, gentleness, ability to minimize honey consumption, honey production, queen color so the beekeeper can easily find her, bees handle dearths, and build-up ability are a few characteristics looked for. In the southeast, the ability to withstand summer heat is another desirable characteristic. Matching the type of bees' traits to the local environment is desirable. For instance, Italian bees do well in the southeast.

Raising queens by grafting is a skilled endeavor. It takes time and resources to raise queens. The queen's development schedule must be adhered to for success with the queen's brood pattern evaluated at the end. It is very important to evaluate the queen's brood pattern to see if you were successful and evaluate your queen's characteristics over several years.

We want bees with superior genetics raised in an excellent environment. Determining a superior queen is often not easy. Selecting one trait may suppress another desirable trait. While walk-away splits should be minimized as much as possible, sometimes it is difficult to do in the southeast until a mated queen is available. One possible solution is to use swarm cells but swarm cells may propagate a colony's propensity to swarm. There are several methods to raise queens with desired traits. The old saying you can make either bees or honey is true.

The first of June, at the end of the nectar flow, is also an excellent time to split colonies. There are a lot of bees available right after the nectar flow and mated queens and queen cells are available. One of the issues is getting the colony to draw-out comb over the summer dearth. Often the colony will need to be put on cotton fields in mid-July or taken to the mountains for the sourwood flow. It should be noted that drawn comb, especially in the brood chamber, improves overwintering success. Drawn comb is required to ensure the cluster is in the bottom of the hive with plenty of honey and pollen going into winter. If the colony does not draw out the comb after a June split, the brood nest will remain in the hive areas where the drawn comb occurs. This may be a feed chamber super or honey super. Feeding sugar syrup rarely enhances drawing out comb. Normally a nectar flow is required. The colony normally needs to have about 80 percent comb utilization before them drawing out more comb.

Figure 8. Festooning chain/drawing out comb. (Photo courtesy: David MacFawn)

After splitting, the split can be moved three to five miles away or left in the original yard. Moving the split works better since the field bees will stay with the moved split. If the split is left in the original yard, an extra frame of bees and brood should be placed in the split. Often it is better to locate the split directly next to or in front of the original location to minimize losing field bees in the split.

A frame of bees of various ages is best to place in both half splits. This will ensure that you do not have all the same age workers and minimizes the need for a lot of nurse bees initially.

Figure 9. A frame of various ages brood. (Photo courtesy: David MacFawn)

Utilizing swarm cells to place in each split half may also be considered. This will shorten your queen development time by about a week or so. However, some beekeepers believe utilizing swarm cells propagates swarm tendencies. It should also be noted that leaving swarm cells in the half with the original queen may result in a swarm being issued from that split half.

Figure 10. Swarm cells on the bottom of a frame. (Photo courtesy: Kathy Carpineto)

There are several ways to avoid swarming. Making splits and utilizing a walk-away split, queen cells, mated queen, or swarm cells all have their benefits. Demaree may also be utilized to simulate a swarm condition in the colony and save all your bees for honey production. Moving the split three to five miles away from the original bee yard works best. Whether you split or not, make sure you get back to what your needs and strategy are. If you need to make increases due to losses, splitting may be the way to go. However, if you want to make a honey crop, then Demaree may be the way to go.

Splitting: A Pictorial Guide

Splitting a colony is one way to control swarming. It is also a way to make increases to make up for lost colonies. When making a split, the new split should contain capped worker brood with some worker eggs and larvae, honey, and pollen. It takes workers consuming honey and pollen to produce worker jelly to feed worker larvae, to ensure worker brood is completely fed. During the spring nectar flow in South Carolina there is a tradeoff between making splits and obtaining a honey crop.

Figure 11. Frame of bees. (David MacFawn)

Enough nurse bees are required to care and cover the brood on cool nights. Older field bees are required to forage for water, pollen, and nectar to feed the young larvae. When making a split, if swarm cells are available, the beekeeper will reduce the time needed for a split to become a functional colony. With a capped swarm or queen cell, it takes an average of about seven days for the queen to emerge from a freshly capped pupa. Add another ten to 14 days for the queen to mate and start laying plus another 19 to 21 days on average for first workers to emerge for a new split. This means it takes approximately 36 to 42 days for new workers to emerge. Add an additional 18 to 21 days for workers to start foraging means it will take approximately seven to nine weeks for foragers to start collecting nectar, pollen, or water.

Most colonies swarm either right before or during the nectar flow. However, if mated, local queens are used the wait time for new foragers can be reduced by approximately 14 to 17 days compared to moving a queen cell (seven days for emergence and seven to ten days for mating). If a walk-away split is made, where the bees raise their own queen, it will take approximately 15 to 17 days for the queen to emerge, ten to 14 days to mate and start laying, 19 to 21 days for workers to emerge in a new split, and an additional 18 to 21 days to become field bees. In total that is approximately nine to ten weeks to produce field bees from a walk-away split. This maturation time will impact honey production. Hence, the old saying you can make either bees or honey is true.

Development Time

Type of split	Queen develops	Emerge / mate / lay	Workers emerge	Total to worker emergence	Field bees	Total to field bees
Mated queen	0	3-7	19- 21	22-28 days	18-21	40-49 days
				3.1-4.0 wks.		5.7- 7.0 wks.
Capped Queen cell	0-7	10-14	19-21	29-42 days	18-21	47-63 days
				4.1-6 wks.		6.7-9 wks.
Walk-away / egg	15-17	10-14	19-21	44-52 days	18-21	62-73 days
				6.3-7.4 wks.		8.9-10.4 wks.

When making a split, the split should be closely monitored for about three weeks to see if you need to add another frame of brood. Bees typically only live an average of five to six weeks during warm weather and it may take six to ten weeks to obtain field bees. Splits should be fed sugar syrup as a safeguard with the understanding that during the nectar flow stored syrup may intermingle with collected nectar. Bees typically will stop using sugar syrup when fresh nectar is available. However, due to the reduced size of a split, and normal population losses, it is wise to offer sugar syrup until they become a functional colony. If another frame of bees and brood is added from a strong non-split colony, and if the nectar flow is one-quarter to one-half complete, the brood that is removed from the strong non-split colony will minimally affect honey production. This is because the time needed for any brood taken from the donor hive has yet to complete its development into mature field bees.

Figure 12. Frame of capped workers with some eggs and larvae on left. (Photo courtesy: David MacFawn)

When selecting a frame of brood, you want mostly capped brood with some eggs and larvae. Capped brood has completed its feeding stage which means the split does not need as many nurse bees. The eggs and larvae are an insurance policy in case the bees need to raise a new queen. Also, you want all different stages of brood so you have a continuous supply of bees. A minimum three-frame split is recommended with a frame of capped brood, eggs and larvae, honey, and pollen. A five-frame split is better with all stages of brood since it will build up quicker than a three-frame split.

Figure 13. Frame of worker brood with eggs and larvae on the top. (Photo courtesy: David MacFawn)

The figure is a good frame for a split. It contains a lot of capped worker brood, with some larvae and eggs on the upper part of the frame. Typically, at least one total deep frame with brood should be used in addition to a frame of honey and a frame of pollen. The brood frame should be placed in the middle of the three frames for warmth. All three frames should be covered with as many bees as possible.

The split should typically be moved more than three miles to retain your field bees. If you decide not to move the split and leave it in the same bee yard, the field bees will return to the original location. Leaving the split in same bee yard will still work, if you have enough nurse bees to cover the brood. The split needs to be monitored closely and another frame of bees and brood added if necessary. The split should be fed sugar syrup.

Figure 14. Eggs and larvae with some capped worker brood. (Photo courtesy: David MacFawn)

The figure has a lot of eggs and larvae for a split without a lot of nurse bees. The eggs and larvae require a lot of nurse bee visits with the resulting large amount of nectar, honey and pollen to feed these larvae. This frame may not be the optimum choice for a split. An optimal frame contains capped brood in addition to larva and eggs. This will allow a continuous supply of bees until a queen starts laying.

Figure 15. Frame of pollen from various sources (different color). (Photo courtesy: David MacFawn)

The figure is an excellent frame of pollen for a split. Note the various pollen colors indicating the pollen is from various sources. This results in a varied pollen diet. It takes brood, honey, pollen, and bees for a split to be successful.

Three or more frames for the split should be placed in the middle of a brood chamber with additional frames on either side. During a nectar flow, the additional frames may contain foundation since the bees will typically draw the cells out. If a split is made after the main nectar flow, frames with drawn comb are preferred. The colony should typically be treated for Varroa prior to the split to ensure the treatment chemicals do not interfere with requeening.

In South Carolina, splits may be done during the spring nectar flow, in June after the nectar flow is over while the colony still has a lot of bees and brood, and in the August time frame. Splitting in August with young local mated queens is good preparation for the next year when queens are not available during the typical spring buildup period.

Figure 16. Frame of worker brood. (Photo courtesy: Kathy Carpineto)

Figure 16 would make a good brood frame for a split. Note: it does not have an insurance policy of additional eggs and larvae. Enough nurse bees and field bees need to be retained to cover the brood on cool days and nights.

Figure 17. Fresh white wax on edges of comb. (Photo courtesy: David MacFawn)

The figure shows fresh white wax and is typically one of three signs of a nectar flow occurring. The other two signs are fresh nectar in the combs and the bees flying with "a sense of purpose"--not languishing as they leave the hive.

Figure 18. Swarm cells on the bottom of a frame. (Photo courtesy: Kathy Carpineto)

Figure 19. Swarm cells protruding from bottom of a frame. (Photo courtesy: Kathy Carpineto)

Figure 20. Swarm cells/cups on the frame bottom with two emergency cells midway up the frame. Note the pollen between the queen cells and the honey in the upper left of the frame. (Photo courtesy: Kathy Carpineto)

This frame was part of a split where the queen did not take for some reason; therefore, the bees built emergency cells. This split should be monitored closely to ensure they have enough bees to get them through until the foragers start maturing. Note the lack of brood on the frame. This split may not have had a laying queen. This split should also be monitored to determine if another frame of brood and bees should be added to sustain the split.

When assessing queen cells, you want larger, more sculptured cells. The more sculpturing the better the cell is considered. Usually Supercedure/ emergency cells are toward the upper part of a frame and not as numerous as swarm cells toward the frame's bottom.

Splitting a strong colony is one way for swarm control and to make increases. A minimum three-frame split with a five-frame is preferred, with a frame of brood/eggs/ larvae, a frame of honey, and a frame of pollen. The split should be monitored at approximately three weeks to determine if additional brood and bees need to be added. In South Carolina, splits may be created at least three different times of the year. In the March through May time period when the spring nectar flow occurs, in June after the spring nectar flow, and in the August time frame. Usually there is a tradeoff between making splits and making a honey crop.

About David MacFawn

David MacFawn (dmacfawn@aol.com) is an Eastern Apiculture Society Master Beekeeper and a North Carolina Master Craftsman beekeeper living in the Columbia, South Carolina, area. He is the author of three books, all published by Outskirts Press:

1. Applied Beekeeping in the United States
2. Beekeeping Tips and Techniques for the Southeast United States, Beekeeping Finance
3. Getting the Best From Your Bees

David has kept bees in Maryland (Dark German bees), Virginia (Italian), North Carolina (Italian), Colorado (Russian), and South Carolina (Italian and Russian Hybrid). He is a North Carolina Master Craftsman Beekeeper (October 16, 1997), Co-Founded the South Carolina Master Beekeeping Program, awarded 1996 & 2020 South Carolina Beekeeper of the Year, assisted Dr. Fell at Virginia Tech in the Virginia Master Beekeeping Program, Incorporated the South Carolina Beekeepers Association as a 501 C 3 Non-Profit Corporation, and published several (over 60) articles in the American Bee Journal, Bee Culture, and Beekeeping: The First Three Years. He currently publishes in Bee Culture. David is a 2021 CIPA EVVY™ Awards Book Second Place Competition Winner.

During the Eastern Apicultural Society meeting in Greenville, SC, July 15-19, 2019 (where he served as Co-Program Chair) he received his Eastern Apiculture Society Master Beekeeper certification.

From July, 2020 to January, 2021 he was a consultant to Bee-Downtown where he identified honeybee to Six Leadership Domains correlations and colony management enhancements.

He also developed, marketed, and supported spreadsheets analyzing financial aspects of the honey and pollination businesses, and beeswax candle production and sales. David has a BS in Electrical Engineering and a Master's in Business

Administration with concentrations in Finance and Operations Research. David was in the computer business for over 30 years and was a Customer Service Program Manager responsible for worldwide support planning, training/education, logistics, call-center support, and professional services at Sun Microsystems and a subset of this at NCR. David was also a Federal Systems Product Manager responsible for new DOD system definition and development at Data General. David resides in the Columbia, South Carolina, USA area and is an active sideline beekeeper.

Ingram Content Group UK Ltd.
Milton Keynes UK
UKHW020952190323
418755UK00004B/17

9 781914 934445